THE ROYAL COMMISSION ON CRIMINAL JUSTICE

The Ability to Challenge
DNA Evidence

by **Beverley Steventon**

LONDON: HMSO

Any views expressed in this report are those of the author(s) and do
not necessarily reflect the views of the Royal Commission

CONTENTS

Foreword

The Royal Commission on Criminal Justice was announced in March 1991. The fact that the Royal Commission was announced shortly after the release of the Birmingham Six suggested that the admissibility and reliability of forensic evidence would be an important area under consideration. DNA profiling evidence is a forensic technique that has attracted considerable attention from scientists, lawyers and the media. It was in response to some of this attention that I examined the terms of reference of the Royal Commission and submitted a research proposal for their consideration. The proposal was accepted and research carried out between November 1991 and March 1992 has resulted in the production of this report for the Royal Commission on Criminal Justice.

The project was completed with the assistance of the Forensic Science Service, the Crown Prosecution Service, the organizations/individuals who carry out DNA profiling on behalf of the defence and the defence lawyers who participated in the study. I wish to express my thanks to all of the above and, in particular, to Dr. Patrick Lincoln of London Hospital Medical College who gave a great deal of his time to assist with this research.

Beverley Steventon
Division of Law
Coventry University

AIM OF THE PROJECT

The aim of the research project was to evaluate the problems which a defence lawyer may face, both pre-trial and at trial, when he is dealing with a case involving DNA profiling evidence.

The project was designed to indicate the problems that defence lawyers may come across in each of the areas below:-

i.   Initial assessment of the DNA profiling evidence.

ii. Decision to obtain expert advice.

iii. Locating and funding a suitable expert.

iv. Disclosure of the DNA profiling results to the defence.

v.   Dealing with DNA profiling evidence in court.

The evaluation of these problems by means of both quantitative and qualitative analysis will be set within the context of the type and number of cases in which DNA evidence has occurred.

DNA PROFILING

The Technique

Prior to consideration of the methodology and results of the research, it is appropriate to give an outline of DNA profiling as a forensic technique. It should however be noted that a discussion of the precise technique and it's validity does not fall within the scope of this project.

DNA[1] profiling is a technique which enables the scientist to compare two biological samples and to determine the likelihood that these samples originated from the same individual. DNA profiling was first developed by Professor Alec Jeffreys[2] who developed the technique of multilocus probing (MLP).[3,4] It was soon realized that the technique had considerable potential with regard to forensic work.[5-7]

The technique has developed over the last few years and there are now a number of different methods of DNA profiling. The main technique currently used by the Forensic Science Service and the Metropolitan Police Laboratory is that of a single locus probing (SLP).[8] The technique of single locus probing is considered more suitable for forensic work.[9] However, it should be noted that these techniques continue to be refined and new techniques of DNA profiling continue to be developed.[10,11]

This continuing state of development of a forensic technique is unique to DNA profiling in the speed with which the changes are occurring. This may pose particular problems for defence lawyers in the finding of an expert who is able to carry out an independent examination of the samples and for the courts in determining when the evidence should be admissible.

Interpretation of the results

DNA profiling evidence appeared to be introduced and accepted by the courts very rapidly[12] both in the UK and other countries. Initial media reports claimed that DNA profiling was the forensic breakthrough of the century.[13] It is reasonable to suggest that at this point in time there may have been a feeling amongst many defence lawyers that frequently DNA profiling evidence was overwhelming and that it was not possible to contest the evidence in court.

Shortly after the introduction of DNA profiling evidence in criminal cases questions were raised in relation to the technique of DNA profiling and in particular to the interpretation of the results. The importance for the defence of an independent analysis of the results was demonstrated by cases such as People v Joseph Castro (1989)[14] and R v Tran (1991),[15] an American and an Australian case respectively. In these cases there was evidence of inadequate laboratory technique and a dispute over the interpretation of the results.

Although the validity of the DNA profiling technique itself appears to be accepted by scientists worldwide,[16] there are still disputes over the interpretation of the results. This is demonstrated by recent articles in scientific journals[17-19] and reports in the media.[20,21] In view of these disputes it is important to ensure that, under the current adversarial system, the defence are able to obtain an independent analysis of DNA profiling evidence. Independent analysis is necessary for the defence both to advise their client and to make an informed decision whether or not to contest the prosecution evidence at the trial. This research aims to determine whether this is particularly difficult for the defence in relation to DNA profiling evidence.

Background Data

DNA Profiling for the Prosecution

The majority of the DNA profiling work carried out for the prosecution in England and Wales is carried out by the Forensic Science Service (FSS)[22] and the Metropolitan Police Forensic Science Laboratory (MPFSL).[23] The Metropolitan Police Forensic Science Laboratory covers the Metropolitan Police area whilst the Forensic Science Services' six regional laboratories cover the rest of England and Wales. In the early days of DNA profiling, some work for the prosecution was carried out by Cellmark Diagnostics, who worked with both the MPFSL and the FSS.

Since the introduction of DNA profiling into criminal casework, until the end of 1991, the MPFSL and FSS carried out DNA profiling in approximately 4,131 cases.

MPFSL	1,709
FSS	2,422

The cases concerned were primarily rape and other sexual assaults but DNA profiling has been carried out in relation to a wide variety of offences e.g. murder, wounding, burglaries, robberies.

The figure for the MPFSL can be broken down by year:-

1989	420
1990	497
1991	792
Total	**1,709**

These figures show that there was a 59 per cent increase in the number of cases in which DNA profiling was carried out between 1990 and 1991. This demonstrates a considerable increase in the use of this technique for forensic purposes.

The above figures represent the total number of cases in which DNA profiling was carried out. It is important to consider that such work does not always result in evidence against the suspect and that a suspect may be exculpated by this technique. This is demonstrated by a breakdown of figures from MPFSL.

Between January and November 1991 the MPFSL carried out DNA profiling in 761 cases with the following results:-

In 40 per cent (304) of these cases profiles matching the suspect were obtained.

In 12 per cent (91) of these cases profiles were obtained which *excluded* the suspect.

In 38 per cent (289) of these cases no profiles were obtained.

In 10 per cent (76) of these cases a profile was obtained which could not be assigned to any suspect.

That DNA profiling may exclude a suspect is of considerable value not only to the suspect but also to the police who may then pursue a different line of enquiry.

In cases where the DNA profile from the scene of the crime matches that of the defendant, the defence may wish to obtain their own independent assessment of the DNA evidence. In order to carry out this assessment the defence may wish to obtain the remains of the scene of crime sample in order to repeat the scientific analysis. The MPFSL and the FSS estimate that in 35 per cent and 30 per cent of crime stains, respectively, there is sufficient material remaining for further analysis. This suggests that in many cases where the defence wish to obtain an independent laboratory analysis of the scene of crime sample they will be unable to do so due to insufficient sample remaining. In these circumstances, the defence may wish to examine the results obtained by the prosecution. This may include visiting the laboratory in order to examine the autoradiographs.

The MPFSL stated that they had never been asked by the defence to supply samples for independent analysis, but in 71 cases there had been defence examinations of the results of DNA profiling and in 17 of these cases this had involved close scrutiny of the autoradiographs.

DNA Profiling for the Defence

There are a number of organizations/individuals within England and Wales who will carry out DNA profiling or evaluate DNA profiling results for the defence. During the course of the research seven such organizations/individuals were contacted by the researcher. Four of these organizations/individuals participated in the study. The four participants in the study had dealt with approximately 125 cases between them. The remaining three organizations/individuals contacted did not wish to participate in the project. Each stated that they had dealt with only a small number of criminal cases involving DNA profiling and this was given as a reason for not wishing to participate. There is one further company who has carried out criminal DNA profiling work in the past but no longer does so.

METHODOLOGY

The principal aim of the research study was to evaluate the problem faced by defence lawyers dealing with cases involving DNA profiling evidence. This would then be set within the context of the type and number of cases in which DNA profiling has occurred.

The principal aim of the project, by definition, necessarily involved canvassing the views of defence lawyers who had dealt with such cases. However the project was not limited to this one approach. It was also felt that the views of forensic scientists who have acted on behalf of defence lawyers in DNA profiling cases were of value in relation to a number of the areas listed on page 1. The method by which the views of defence solicitors and forensic scientists who act on behalf of the defence were obtained are given in detail below, followed by details of the background information sought.

Defence Lawyers

Defence lawyers who deal with a case involving DNA evidence will consider it and then make an initial decision as to whether or not to obtain their own evaluation of the DNA profiling evidence. The researcher wished to obtain the views of both defence lawyers who did, and those who did not seek an independent evaluation of the DNA profiling evidence. The views of the first in respect of the problems encountered and of the latter in respect of their reasons for feeling that such an evaluation was not necessary.

Selection of the sample

It was recognized at the outset of the project that the sample available would depend on the willingness of forensic scientists and solicitors to participate in the project.

To obtain a sample of defence solicitors who had obtained an independent evaluation of DNA evidence, forensic scientists who act on behalf of the defence in relation to DNA evidence were approached. In order to protect client confidentiality the forensic scientists were then asked to distribute a letter to solicitors for whom they had carried out DNA work. This letter requested their co-operation with the project. It was felt that the forensic scientists would only participate where this indirect approach was taken. The disadvantage of such an approach is that it is very time consuming.

To obtain a sample of defence solicitors who had dealt with a case involving DNA evidence but who may not have obtained an independent analysis, it was necessary to approach the Forensic Science Service (FSS) and the Crown Prosecution Service (CPS). The FSS provided the names of cases where DNA profiling had been carried out for the prosecution. The names and addresses of the defence solicitors who had acted in these cases were located from CPS files. The defence solicitors were then contacted by the researcher.

In addition to the above, a letter was sent to all the members of the London Criminal Courts Solicitors Association and a letter placed in the Law Society's Guardian Gazette.[24] These letters described the project and invited any member of the profession with an interest in DNA evidence to contact the researcher for a questionnaire.

The Views of the Defence Lawyers

In order to elicit the views of the sample of defence solicitors, each solicitor was sent a questionnaire. For practical reasons the researcher decided that the most appropriate way to obtain the maximum information was to design a detailed questionnaire. This method was preferred to the initial proposal of a brief questionnaire to all solicitors followed by a small number of interviews.

The Questionnaire for Defence Lawyers

A copy of the questionnaire sent to the defence solicitors is to be found in Appendix 1.

The questionnaire did not require the respondents to identify themselves or the name of the case that they were referring to. This strict confidentiality was felt to be essential in persuading defence lawyers to respond, despite the limitations it would place on following up the questionnaires.

Section A – Background to the case

This section of the questionnaire was designed to provide information on:–

    i.   The person completing the questionnaire i.e. the capacity in which the respondent was involved in the case and whether or not this was the first case involving DNA evidence that they had dealt with.

and:–

> ii. The background of the case i.e. offence type, year the offence was committed, plea given by the defendant and the outcome of the case.

Section B – Disclosure of DNA Profiling Evidence

This section of the questionnaire aimed to determine who notified the defence of the existence of DNA evidence, and in particular, whether there was sufficient time pre-trial for the defence to obtain independent analysis of the DNA evidence.

Section C – Expert Assessment of the DNA Evidence for the Defence

The questions in this section were designed to determine whether or not the defence wished to obtain an independent assessment of the DNA evidence. Those defence solicitors who did not wish to obtain such assessments are asked to state why. Those defence solicitors who did wish for such an assessment are questioned on the way in which they located their expert, the work he carried out and the funding of the expert.

Section D – Results of the Defence Expert Assessment

This section aimed to determine the proportion of cases in which the defence experts analysis differed from that of the prosecution's scientist.

Section E – Challenging the DNA Evidence in Court

This section of the questionnaire considers the way in which defence lawyers dealt with the DNA evidence in court. Questions relate to the assistance provided by an expert witness and the funding of that expert witness. This section also allows for the respondent to state any particular problem that they experienced in relation to the DNA evidence.

Forensic Scientists acting for the Defence

To obtain the sample of appropriate defence lawyers described above, a number of forensic scientists who act for the defence had already been contacted. A questionnaire was designed to examine the views of these scientists.

The Questionnaire to the Forensic Scientists

A copy of the questionnaire is to be found in Appendix 2.

The questionnaire did not ask the forensic scientist to identify the name of the case. This was done in order to avoid the problem of client confidentiality. This questionnaire was designed to complement that sent to the defence lawyers.

The questionnaire dealt with the following issues:–

i.   Questions 1–3 – the background of the case i.e. offence type, year the offence was committed and how the defence solicitor located the respondent as a suitable expert.

ii. Questions 4–5 – notification by the defence i.e. was there sufficient time, pre-trial, for the expert to carry out all the procedures that he felt to be necessary for the defence?

iii. Questions 6–7 – the funding of the work.

iv. Question 8 – the nature of the scientific work carried out on behalf of the defence.

v.   Question 9 – did the conclusions drawn by the respondent differ from those given by the prosecution scientist.

vi. Question 10 – the role the expert felt he played in court.

Response rate to the questionnaires

Defence Lawyers

In order to identify the appropriate defence lawyers to whom questionnaires could be sent, organisations/individuals who have carried out DNA forensic work on behalf of the defence were contacted. Four of these were willing to send out letters or provide an address for clients for whom they had carried out DNA defence work. The parties are listed below in alphabetical order. For reasons of confidentiality in a commercial market no indication will be made concerning the number of cases each party has dealt with.

i.   Cellmark Diagnostics

ii. Dr Patrick Lincoln of London Hospital Medical College

iii. The Forensic Science Service

iv. UK Forensic Science Service

The organisations/individuals listed above sent out a total of approximately 125 letters to defence solicitors who had obtained an independent evaluation of DNA evidence. These letters requested that any solicitor willing to participate in the study contact the researcher. The researcher received 76 requests for questionnaires as a result of these letters.

Details of 46 defence solicitors were provided by the CPS from cases names given to them by the Forensic Science Service. These cases were primarily cases which occurred in 1991. Questionnaires were sent out to all of these solicitors asking for their assistance with the study.

6 requests for questionnaires were received from defence lawyers who became aware of the project by other means.

A total of 128 questionnaires were thus sent out to defence lawyers. Out of the 128 questionnaires sent out, 54 were returned. This is a response rate of 42 per cent.

Forensic Scientists

Two forensic scientists who participated in the earlier part of the study, by providing contact with defence solicitors, completed a questionnaire in relation to each of the cases involving DNA evidence in which they had acted for the defence. This questionnaire was complementary to that designed for the defence lawyers.

A total of 58 completed questionnaires were received from the forensic scientists.

Limitations of the study

The research study involved access to confidential material. It was clear from the outset of the project that in order to obtain the assistance of both the defence lawyers and the forensic scientists the respondents could not be asked to identify the individual cases to which the completed questionnaires referred. This clearly placed severe limitations on the follow up of questionnaire responses.

The defence lawyers were treated as one group of respondents irrespective of the means by which they were initially contacted i.e. via letters sent out from forensic scientists or via addresses provided by the CPS. Of the 54 questionnaires returned 46 were suitable for analysis. 34 of these involved cases in which the defence lawyer had obtained an independent assessment of the DNA evidence and 12 involved cases in which the defence lawyer had not wished to obtain an independent assessment of the DNA. The researcher is aware that a number of potential respondents involved with cases where the defence had not wished to obtain an independent assessment of the DNA evidence felt that the information they could provide would be limited and were unwilling to complete the questionnaire. This may be one factor contributing to the low number of respondents in this particular category. However it must also be noted that the means by which the defence lawyers were contact biased the study in favour of those who had obtained an independent assessment of DNA evidence.

The research is based on a relatively small number of respondents and should be viewed essentially as an exploratory study.

The overall results from the two groups of respondents, defence lawyers and forensic scientists, were compared to determine whether the problems experienced were of the same nature, or whether due to the different perspectives from which the two groups were looking at the evidence, the problems differed. However, no attempt could be made to match the responses of lawyers and forensic scientists on a case by case basis as neither group were asked to identify the case to which the completed questionnaire referred. This was unfortunate but unavoidable if the co-operation of the two groups was to be obtained.

RESULTS

Questionnaires to the Defence Lawyers

Method of Analysis

Of the 54 questionnaires returned 46 were subjected to computer analysis by means of SPSS.

The 8 questionnaires not analysed were omitted for the following reasons:-

i.   4 questionnaires related to Scottish cases and these were omitted as it was decided to restrict the study to cases within England and Wales.

ii. 3 questionnaires contained insufficient data.

iii. 1 questionnaire was a duplicate of another questionnaire entered into the study.

Analysis of the Questionnaire to the Defence Lawyers

Each questionnaire was completed in relation to a case involving DNA profiling evidence that the defence lawyer had dealt with. Sections A and B of the questionnaire were completed by all respondents, and hence includes both solicitors who did and those who did not obtain an independent assessment of the DNA evidence.

Section A – Background to the case

The respondents to this questionnaire were primarily solicitors who had acted for the defence in a case involving DNA evidence. 44 out of 46 questionnaires (95 per cent) were completed by solicitors. The remaining 2 were completed by barristers. The reason for this is that the method by which the sample of potential respondents were located resulted in this questionnaire being primarily directed at solicitors.

For 25 respondents (54 per cent) the case in relation to which they were completing the questionnaire was the first case they had dealt with involving DNA evidence. 21 respondents (46 per cent) had dealt with a case involving DNA evidence prior to the case in relation to which they were completing the questionnaire.

The year in which the offences were committed is depicted in Table 1 below.

Table 1: Year the alleged offence was committed

Year	Number of Cases	
1986	1	(2.2%)
1987	2	(4.3%)
1988	6	(13.0%)
1989	7	(15.2%)
1990	11	(23.9%)
1991	17	(37.0%)
MISSING	2	(4.3%)
TOTAL	26	(100%)

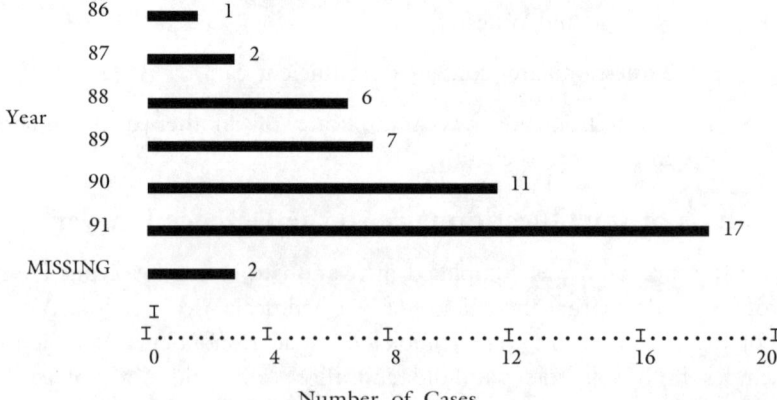

A total of 28 cases (61 per cent) dealt with in this questionnaire occurred in the years 1990 and 1991. This is representative of the rapid increase in the number of cases in which DNA profiling has been used since its introduction into criminal casework. However it should also be noted that the sample selection process may have biased the results towards cases in 1991 see page 11 above.

The offence types covered by the cases dealt with in this questionnaire are depicted by the pie chart Fig. 1.

The most common category of offence for which DNA profiling has been used is sexual offences, in particular rape and attempted rape. This was to be expected as it is particularly likely in such offences that the offender will leave a sample of body fluid, e.g. semen in a rape, at the scene of the crime. Hence DNA profiling is of particular value in such offences. DNA profiling is a technique which is time consuming and expensive and it is therefore likely that it will primarily be carried out in relation to serious offences.

Figure 1

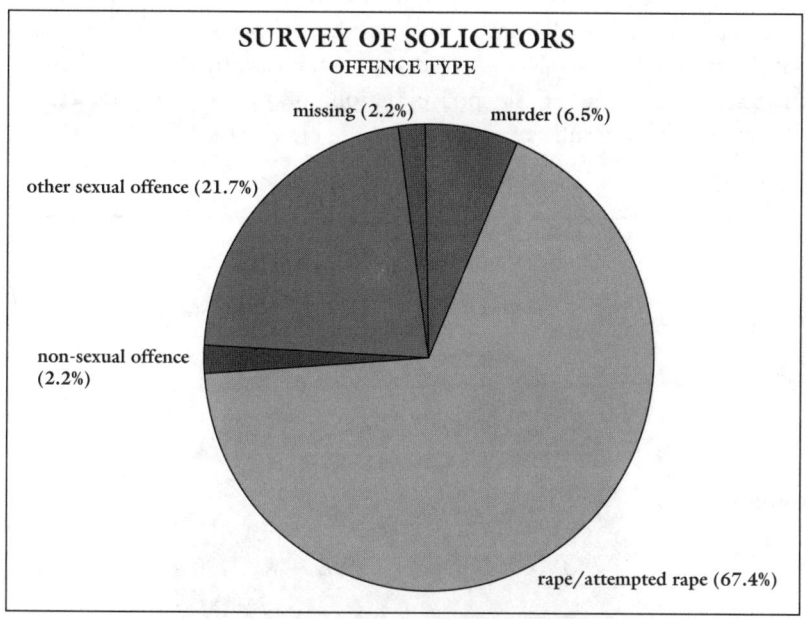

In 36 cases (78 per cent) the defendant pleaded not guilty to the offence charged. In 8 of these cases the defendant was acquitted, in 20 of these cases the defendant was found guilty as charged and in 3 of these cases the defendant was found guilty on some but not all of the counts charged. In 5 of these cases the outcome of the trial was not known. In 1 case the conviction was quashed on appeal and in 2 cases there are appeals pending. It is known that in 1 of these cases the appeal pending concerns DNA evidence.

In 8 cases (17 per cent) the defendant pleaded guilty to the offence charged.

In 2 cases the defendant pleaded guilty to some charges and not guilty to others.

It is not possible from this study to determine whether the DNA evidence was a significant factor in the jury's decision to convict. However it is worth noting that in only 8 of the 36 cases (22 per cent) where the defendant pleaded not guilty on all counts, was he subsequently acquitted.

Section B – Disclosure of DNA Profiling Evidence

The defence were primarily notified of the existence of DNA profiling by the prosecution. In a number of cases the respondents specified whether they were notified by the police or CPS, other respondents simply stated that they were notified by the prosecution. In a total of 39 cases (84 per cent) notification was in this manner i.e. police, CPS or prosecution (unspecified). In a small number of cases the defence were alerted to the possibility of there being DNA evidence in the case by the defendant or during the interview at the police station, or by the forensic science laboratory. These results are depicted by the pie chart in Fig. 2.

Figure 2

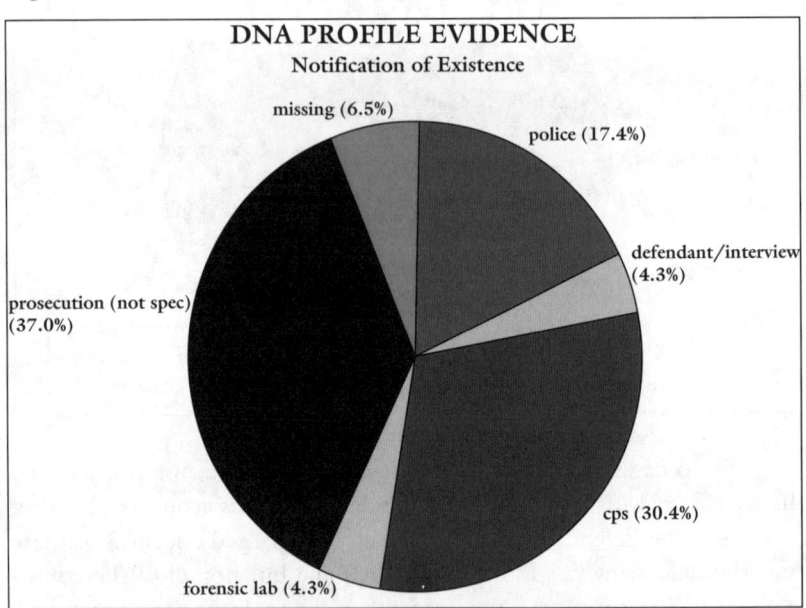

The length of pre-trial notice that the defence had of the existence of DNA evidence in the case varied enormously from approximately 1 year's notice to notification at the actual trial.

The important factor to consider here is that DNA profiling may take many weeks to carry out and so the defence will require a considerable period of time in order to locate a suitable expert and for that expert to carry out the required work.

In at least 8 cases (17 per cent) the defence were notified less that 1 month pre-trial of the existence of DNA evidence.

The respondents were asked whether they were notified in sufficient time pre-trial to obtain an independent assessment of the DNA evidence had they wished to do so.

In 32 cases (70 per cent) the respondents felt that there had been sufficient time pre-trial to obtain an independent assessment of the DNA evidence. In 10 cases (22 per cent) the respondents felt there had not been sufficient time. There were 4 missing values to this question.

In relation to this question comments were invited from the respondents. 4 respondents stated that although there was time for the defence work to be carried out there was a considerable rush. 1 respondent felt that the prosecution had deliberately delayed informing the defence of the existence of DNA evidence in the case. In 5 cases where the defence had not had time pre-trial for the DNA assessment the trial was adjourned to allow for the work to be carried out.

Section C – Expert Assessment of the DNA Evidence for the Defence

This part of the questionnaire divided the 46 respondents into two groups, those who had obtained an expert assessment of the DNA and those who had not.

34 of the 46 respondents (74 per cent) to the questionnaire had obtained an independent assessment of the DNA evidence. Replies were received from only 12 respondents (26 per cent) in relation to cases where the defence had not obtained an independent assessment of the DNA evidence. The small number of respondents to this latter category has been discussed above.

Respondents who did not obtain an Independent Assessment of the DNA Evidence

The respondents who did not obtain an independent analysis of the DNA evidence were asked why this decision was taken. The following refers to the 12 respondents who replied to this effect.

In 7 cases the defence did not feel it was necessary to obtain an expert assessment of the DNA evidence as an alternative explanation was offered for the presence of the defendants DNA at the scene of the crime e.g. the defendant admitted his presence at the scene but denied the offence. The most common way this occurs is the defendant claiming consent to an allegation of rape.

In 3 cases the defendant had decided to plead guilty prior to being aware that there was DNA evidence against him. It should be noted that this category would include persons from whom the police request samples for DNA analysis. The defendant may have decided to plead guilty on becoming aware of the potentially incriminating nature of the samples he has given prior to the results being obtained.

In 1 case the defendant decided to plead guilty on becoming aware of the DNA evidence against him.

In 1 case the DNA evidence obtained by the prosecution did not implicate the defendant and therefore an independent assessment of the DNA evidence by the defence was unnecessary.

Respondents who obtained an Independent Assessment of the DNA Evidence

The following results refer to the 34 respondents who obtained an independent assessment of the DNA evidence.

Legal Aid

26 respondents (77 per cent) applied for Legal Aid to cover the cost of an independent assessment of the DNA evidence. 25 respondents (74 per cent) had Legal Aid granted. Therefore in only 1 case in the sample was Legal Aid applied for and the application refused. It is worthy to note that this case occurred in 1987 and may therefore have been one of the earlier cases involving DNA evidence to be considered by the Legal Aid Board. 1 respondent who applied for and was granted Legal Aid was allowed a maximum of £600. 1 respondent who did not apply for Legal Aid stated that Legal Aid was not applied for due to insufficient time.

Locating a Suitable Expert

Persons considered by the defence to be a suitable expert were located in a variety of ways. These are depicted by the pie chart in Fig. 3.

The most common way for a defence solicitor to locate a suitable expert was to be advised by defence counsel. 8 respondents (24 per cent) located the expert in this way. Other common methods were to have been in contact with the expert before or to be advised by other forensic experts who either do not work in the DNA field or who were unable to carry out the work for other reasons.

The variety of responses to this question suggests there is not an obvious register of appropriate forensic experts for defence lawyers to consult when faced with a case involving DNA evidence for the first time. That such lawyers are unsure where to go for advice is highlighted by the fact that twice during the duration of the project the researcher was contacted by defence solicitors for advice on how to obtain an independent assessment of DNA evidence. One of these enquiries related to a method of DNA typing only recently introduced into criminal casework.[25]

Figure 3

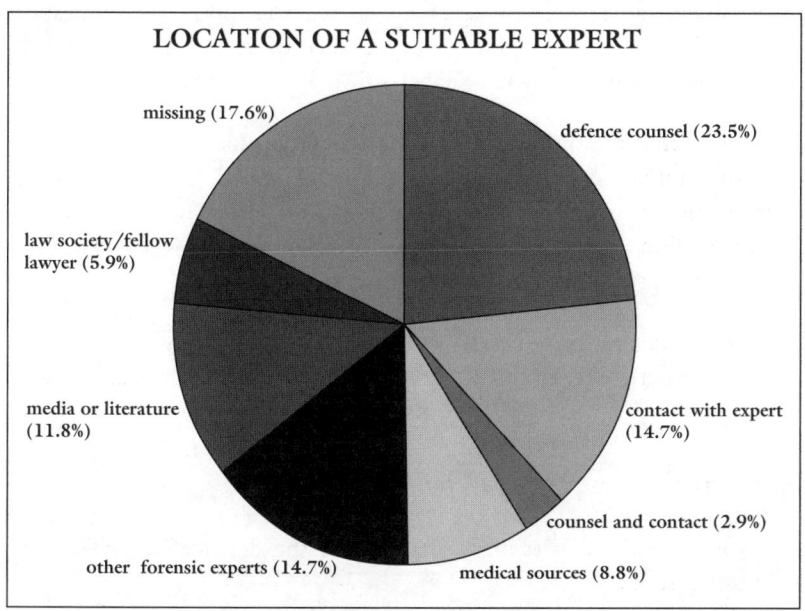

LOCATION OF A SUITABLE EXPERT

missing (17.6%)

defence counsel (23.5%)

law society/fellow lawyer (5.9%)

media or literature (11.8%)

contact with expert (14.7%)

counsel and contact (2.9%)

other forensic experts (14.7%)

medical sources (8.8%)

All 34 respondents to this part of the questionnaire were able to locate and approach at least 1 expert for an assessment of the DNA evidence.

The number of experts approached by the respondents is depicted in Table 2.

Table 2

Number of Experts Approached	Proportion of Respondents
1	25 (73.5%)
2	4 (11.8%)
3	4 (11.8%)
4+	1 (2.9%)
Total Respondents	34 (100%)

A total of 9 respondents (27 per cent) contacted more than 1 expert. Of those who contacted more than 1 expert, 3 did so because the first expert approached was unable to carry out the work either because of their workload or because of a lack of experience in the field of forensic DNA profiling.

It is interesting to note that 2 respondents approached more than 1 expert because the experts approached were experts in different fields.

DNA profiling involves the fields of molecular biology, population genetics and statistics. There is some evidence to support the suggestion that the defence may stand a greater chance of successfully contesting DNA evidence where they have experts from each of these fields.[26]

A further reason for the defence consulting more than 1 expert is that the initial expert approached agrees with the prosecutions interpretation of the evidence. The defence may then wish to obtain a second opinion in the hope that it is more favourable to their case. It is known that this occurred in 1 case in the study.

Although all respondents to this part of the study were able to locate an expert, 3 respondents commented that they had difficulty in locating an expert.

Work carried out for the Defence by the Expert

The work that the expert may carry out for the defence falls into three main categories:–

i. Where the prosecution have analyzed a sample from the scene of the crime and a sample from the defendant, the defence expert may wish to obtain the remains of the sample from the scene of the crime and repeat the scientific analysis carried out by the prosecution. It is reasonable to argue that this is the only way that a full independent assessment of the evidence can be carried out. The problem for the defence is that the sample from the scene of the crime is often small and degraded. Indeed, after the prosecution have carried out DNA profiling there may be no sample remaining. Estimates from the MPFSL and FSS suggest that this will occur in 65–70 per cent of cases.[27] Despite these difficulties 15 respondents (44 per cent) who had obtained expert evidence stated that their experts had repeated the investigations carried out by the prosecution. This proportion seems comparatively high and it should be taken into account that the respondents to this questionnaire are not scientists and may not have been aware of the precise nature of the work carried out by their expert.

ii. 8 respondents (24 per cent) who obtained expert evidence stated that their expert had carried out work which differed to that carried out by the prosecution. One reason for this is that the defendant may provide samples to the defence expert which he is not willing to supply to the prosecution and which cannot be taken without his consent.

iii. 24 respondents (71 per cent) who obtained expert evidence stated that their expert had carried out an evaluation of the results obtained by the prosecution. This necessitates the defence expert having access to the working notes, the results, the electrophoretic gels[28] and the autoradiographs[29] produced by the prosecution. This method of assessment was the most common method used by defence experts in the study. This process does not require any independent scientific analysis of samples and is therefore the quickest and simplest way of assessing the prosecution's results. It is valid to argue that this re-examination of the prosecution's results cannot amount to a full independent assessment of the evidence. However even when a defence expert wishes to carry out an independent laboratory analysis he will not be able to do so where all the relevant scene of crime sample has been used up. In 13 cases (38 per cent) the expert used this method alone to evaluate the evidence against the defendant, whereas in 11 cases the experts also carried out their own analysis of the samples.

Section D – Results of the Defence Expert's Assessment

The results below refer to the 34 respondents who obtained expert assessment of the DNA.

22 respondents reported that their defence expert's assessment of the DNA supported the conclusions of the prosecution expert in all respects.

13 respondents reported that their expert's assessment of the DNA evidence differed from the assessment made by the prosecution expert.

In 3 cases the defence had obtained advice from more than 1 expert and 1 expert had agreed with the prosecutions results whilst the other had disagreed. These 3 cases therefore fall within both of the above categories hence the total is greater than 100 per cent. The defence is free to approach any number of experts for evaluation of the prosecutions results and need not indicate to the court the reports from all of the experts that they had consulted. They are free to present only the evidence that supports their case.

2 respondents did not reply to this question.

Of the 13 respondents who received an expert assessment of the DNA which differed from the assessment made by the prosecution expert, in 1 case the conclusions drawn by the defence scientist implicated the defendant to a greater extent than the conclusions drawn by the prosecution. The defence in such circumstances are under no obligation to draw the courts attention to these results.

In the remaining 12 cases the opinion drawn by the defence expert was more favourable to the defendant than the opinion drawn by the prosecution expert. In 3 of these 12 cases the respondents stated that the difference in opinion between the experts was minor. The other 9 respondents did not comment on the significance of the difference of opinion between the experts, though comments were made regarding the nature of the difference.

Where the respondents had stated that their experts conclusions differed from those drawn by prosecution they were invited to comment.

Prior to giving a summary of these comments it is necessary to indicate the likely areas of dispute between experts comparing DNA profiling evidence. In order for the forensic scientist to determine the likelihood that the sample from the scene of the crime has originated from the defendant a DNA profile obtained from a sample from the defendant will be compared with a DNA profile obtained from the scene of crime sample e.g., a vaginal swab from the victim of an alleged rape. In multilocus and single locus DNA profiling the scientist will first determine whether or not the bands in the two profiles he is comparing match. The scientist will then calculate statistically the chance of a person at random from the population having a profile which matches the crime profile.

In order to calculate these statistics the scientist uses a database which contains the frequency with which a given band will occur in the population.[30] It is in these areas of band matching, database frequencies and statistical calculations that the discrepancies between the defence and prosecution experts occurred. It is difficult to assess the potential effect of these discrepancies on the jury and hence the significance of the difference of opinion between the experts.

The comments made by the 13 respondents whose expert evaluations differed from those drawn by the prosecution are summarized below:–

6 respondents stated that there was a dispute over the statistical calculations.

3 respondents said that their expert had disagreed over the matching of bands between DNA profiles.

2 respondents stated that the defence expert had expressed concern over possible contamination of the scene of crime sample and/or degradation of the sample.

2 respondents stated that there was concern over the scientific process of DNA profiling.

Disputes over the interpretation of the results of DNA profiling between defence and prosecution experts is reflected by the out of court arguments of scientists which are currently being expressed in the scientific journals.[31, 32]

Section E – Challenging the DNA Evidence in Court

In this section the word 'challenge' has been interpreted to include cross-examination of the prosecution expert as well as the calling of the defence expert to testify in court.

22 respondents (65 per cent) contested the prosecutions DNA evidence in court. It was considered that the defence may have wished their expert to assist them in three ways. The defence may have obtained information from their expert in order to enable them to cross-examine the prosecution expert. The defence may have wanted their expert to attend court in order to advise them on cross-examination of the prosecution expert or the defence may have wanted their expert to attend court in order to give evidence for the defence. Respondents were asked to say what assistance they required from their experts in challenging DNA evidence in court. They were free to reply to more than one category e.g. the defence may have required the expert to attend court in order to advise them on cross examination of the prosecution expert and to give evidence on their behalf. Therefore the percentage sum is greater than 100 per cent.

Of the 22 cases where the defence wished to challenge the DNA evidence:–

In 18 cases the defence expert provided information which assisted the defence in their cross-examination of the prosecution expert witness.

In 12 cases the defence experts attended court in order to advise on cross-examination of the prosecution expert witness.

In 5 cases the defence wanted their expert to attend court with a view to calling the expert to give evidence on their behalf. In 3 of these cases the experts did attend court and were called to testify. In all 3 cases where the expert was called to give evidence the respondents had stated in the questionnaire that their expert disagreed with the prosecution conclusions on the DNA evidence.

Table 3 below shows a cross-tabulation analysis of the defence lawyers who wished to challenge the DNA evidence in court and the conclusions drawn by the defence expert.

Table 3

	Did you wish to challenge the DNA evidence in court?		TOTAL
The expert analysis	YES	NO	
supported the prosecution's conclusions in all respects	13	9	22
deviated from the prosecution's conclusions	11	2	13
TOTAL	24	11	35

The table shows that in 11 cases where the defence employed an expert whose conclusions differed from those of the prosecution expert the defence wished to challenge the evidence in court. There were 2 cases where the defence experts conclusions differed from those drawn by the prosecution expert and yet the defence lawyer did not wish to challenge the evidence in court. In 1 of these cases the deviation was insignificant and in 1 case the defence experts report implicated the defendant to a greater degree.

In 13 cases where the defence and prosecution experts agreed on the conclusions to be drawn from the DNA evidence, the defence still wished to challenge the evidence in court. (In 3 of these cases the defence had an alternative expert whose opinion did differ to that drawn by the prosecution). For the defence this is an advantage given by the adversarial system. The defence have the right to challenge the prosecutions forensic evidence in court in front of the jury even where the report provided by the defence expert supports the conclusions drawn by the prosecution expert. In such circumstances the defence are unlikely to call their expert as a witness but they may wish to use his advice to enable them to effectively cross-examine the prosecution expert.

Assistance Provided by the Expert

32 respondents (94 per cent) who consulted an expert felt that they had been assisted by the expert in evaluating the evidence. This assistance enabled the defence to advise their client and/or assisted them in court in the defence of their client.

The 34 respondents who had consulted an expert were invited to comment on the way in which they felt the expert had assisted them.

27 respondents (79 per cent) chose to comment. Those respondents who commented felt that they had been assisted by the expert in one or more of the ways indicated below:-

i.   Understanding the technique of DNA profiling.

ii.   Evaluating the case.

iii. Advising the client.

iv.   Cross-examining the expert for the prosecution.

The responses fell into two main categories. 15 respondents (44 per cent) felt that their expert had assisted them in evaluating the case and in enabling them to advise their client.

12 respondents (35 per cent) felt that their expert had assisted in their understanding of the technique of DNA profiling and had assisted their cross-examination of the prosecution expert.

This categorization of the responses gives an indication of the way in which the respondents felt they had been assisted by their expert.

3 respondents stated that as a result of the defence experts conclusions their client decided to plead guilty.

Problems Experienced in Relation to the DNA Evidence

All participants in the study (46) were invited to comment on any specific problems they had experienced in relation to the DNA evidence in the case that they had dealt with.

The comments made by the respondents were divided initially into four broad categories:–

i.   Respondents who sought an independent assessment and experienced problems concerning the DNA evidence.

ii.   Respondents who sought an independent assessment and had no problems concerning the DNA evidence.

iii. Cases where the DNA evidence was not in issue and the respondent did not seek independent assessment of the DNA evidence.

iv.   Missing values – the respondent made no comment or made a comment which did not fall into one of the above categories.

Respondents who sought an independent assessment and who experienced problems concerning the DNA evidence

16 of the 34 participants (47 per cent) in the study who sought an independent assessment of the DNA evidence gave comments to the effect that they had experienced problems directly related to the DNA evidence.

These comments are central to the aim of the project and are worthy of reproduction in full. The comments are reproduced below without any alteration, however for reasons of confidentiality the comments are numbered and are not linked to any respondent:–

i. 'Insufficient literature to understand what it means initially and to advise client appropriately prior to discussing position with expert'.

ii. 'Without expert assistance, we were unable to elicit the necessary information. DNA results meant little without expert explanation'.

iii. 'Comprehension of techniques and terminology'.

iv. 'It is horrendously difficult to understand. All experts now require to be paid within 30 days of completing their reports and or giving evidence and the Central Taxing Teams don't then pay for some 8 to 12 months. There are also problems actually finding an expert. Working out what everyone is saying and what it means'.

v. 'Getting financial approval for expert in time – only possible due to court postponing trial – in this case the defendant was in custody for other matters and was not prejudiced – if he had not been in custody for those matters, that delay would have been at his expense as bail was refused.

Without our expert reviewing prosecution evidence we would have had to take their evidence completely on trust'.

vi. 'The fact that all samples were 'used up' in prosecution tests'.

vii. 'We often find that after the prosecution have done their own tests, there is too little staining etc. left on the exhibits for our own tests to be carried out to our satisfaction and our expert has to use results already obtained by the prosecution'.

viii. 'We understand that for copyright reasons the defence expert was unable to use the same probes as the prosecution expert and was accordingly unable to exactly replicate the tests that they had carried out. The Writer's experience has been that requests to the Law Society/Legal Aid Board for prior approval to instruct experts are usually met with a negative response and it is left to the solicitor to take a risk on taxation. In this case the Crown Court disallowed the defence expert's fees in the first instance and we had to make representations to the Taxing

26

Officer which fortunately were successful otherwise we would have been some £1,132.80 out of pocket!'.

ix. 'There are very few experts and those that do exist are seriously overworked. Materials are not available to the defence to carry out their own independent tests. The results of the prosecution tests are not made known until shortly before trial. There is always a delay in obtaining instructions on whether to instruct an expert particularly if the accused is in custody'.

In the current climate police forensic evidence cannot be trusted.

If the prosecution were intending to rely on DNA evidence the defence should be invited to instruct an expert who can double check the findings of the police forensic scientists at the time the tests are carried out and thus an agreed report can be put to the court in the majority of cases. The costs of this should be automatically covered by Legal Aid where an alibi warning has been given or some other form of defence justifying checking DNA sampling has been put forward'.

x. 'Not enough experts nor experts who are not paid by large corporations and also might be willing to pooh-pooh potency of DNA evidence. (We could be labouring under a paranoid delusion on this but we wern't aware of any reputable subversive experts being available)'.

xi. 'The two experts were opposed in their scientific findings.

Reluctance of Legal Aid Board/Court to allow expert to challenge DNA evidence'.

xii. 'Difficulty in obtaining 'original' bar codings from prosecution procedure could have been explained in simpler detail and with visual aids to the jury'.

xiii. 'Prosecution disclosure would add prosecution not competently conducted'.

xiv. 'Prosecution were not keen to help re disclosure. In my view no DNA which is challenged by the defendant should be admitted without full challenge by the defence with expert help'.

xv. 'I didn't find the defence report helped. It didn't analyze prosecutions report or give any help as to lines of attack or questions to ask. It seemed to be treated as a purely scientific enquiry with little attention to adversarial process. There were no references to recent literature querying the perfection of DNA profiling'.

xvi. 'Very difficult and frustrating finding experts prepared to criticise the whole DNA evidence approach both in its presentation and its calculations'.

The most frequent problems faced by the respondents can be summarized as below:-

4 respondents expressed concern over the difficulty in understanding a scientific technique as complex as DNA profiling and the value of employing a defence expert to assist in this respect.

4 respondents expressed concern over the problems faced by the defence if they wished to repeat tests carried out by the prosecution. Such problems may occur where the tests carried out by the prosecution have utilized all the available sample (in the case of a scene of crime sample this will be irreplaceable) or where the exact materials e.g. DNA probes used by the prosecution are not available to the defence for commercial/ copyright reasons.

3 respondents commented on the difficulties of finding an expert. Experts who are suitably qualified to carry out DNA forensic work for the defence are small in number and may therefore be in considerable demand.

4 respondents included comments related to the cost of DNA profiling. 2 responses related to obtaining Legal Aid to cover the cost of analysis pre-trial, 1 in relation to a negative response by the Legal Aid Board and 1 in relation to delay in obtaining approval from the Legal Aid Board. The other 2 responses concerned costs incurred by the attendance of the expert at court. Such costs come out of central funds. You cannot obtain prior authority for such costs and the solicitor takes a risk as to whether or not such costs will be met. These costs authorized by Crown Court may be subject to taxation and payment may not be for a period of several months.

3 respondents reported problems with disclosure of the DNA evidence by the prosecution.

Respondents who sought an independent assessment and experienced no problems in relation to the DNA Evidence

14 of the 34 (41 per cent) who sought an independent assessment of the DNA evidence respondents stated that they experienced no particular problems in relation to the DNA evidence in this case.

DNA Evidence was not in issue in the case and the respondents did not seek an independent assessment of the DNA evidence

The 12 respondents in the study who had not wished to obtain an independent assessment of the DNA evidence did not reply to this question. In these 12 cases the DNA evidence was not in issue in the case e.g. where the defendant claimed consent to an accusation of rape. The logical assumption from that is that the DNA evidence would not have raised any problems in these cases. However 1 respondent from this category made some general comments at the end of the questionnaire concerning DNA evidence. These comments are reproduced below:-

xvii. 'General feeling that DNA evidence is impossible to challenge. I have been surprised to find how often no results can be obtained by DNA profiling.

Publication of a reliable list of experts with fees etc would be helpful. Legal aid is always a problem in particular experts do not want to have to wait months for their fees.

DNA evidence is always served very late. Often the prosecution are blasé about this and we have to apply to keep cases out of lists until it is available. Even if intercourse is conceded we want to know all the facts'.

Missing Values

4 respondents chose not to reply directly to this question. However 2 of these respondents made comments concerning the DNA evidence in the case and these are reproduced below:–

xviii. 'Although the defendant was convicted we quite clearly showed that the prosecution forensic expert had been rushed by the police and his calculations were incorrect.

Also the defendant, although European, did not originate from the U.K. and therefore the use of the U.K. database was wholly inappropriate, and could have dramatically altered the final statistics'.

The above comments suggest that this respondent had no particular problems in relation to the DNA evidence in the case. However, it was felt that in court the defence were able to demonstrate errors in the conclusions drawn by the prosecution from the DNA evidence.

xix. 'The forensic evidence all pointed to our client but the DNA did not. Having made 4 statements to say rapist ejaculated the victim changed her mind after being told DNA negative regards the defendant!'.

In this instance the DNA evidence produced by the prosecution did not implicate the defendant. However, it would appear from these comments that the DNA results were negative with regard to the defendant and so although they did not implicate the client they did not exculpate him either. The case was proceeded with on the basis of other evidence against the defendant. In a case such as this it is very important that all the prosecution results are disclosed to the defence i.e. negative as well as positive to enable the defence to fully evaluate the case against their client.

Questionnaire to Forensic Scientists
Method of Analysis

Out of the 58 questionnaires returned, 49 were subjected to computer analysis by means of SPSS.

The 9 questionnaires not analyzed were omitted for the following reasons:–

i. 4 questionnaires related to Scottish cases and 1 to a Hong Kong case. These were omitted as it was decided to restrict the study to cases within England and Wales.

ii. 4 questionnaires contained insufficient data.

Analysis of the Questionnaire to Forensic Scientists

Each questionnaire was completed in relation to a case involving DNA profiling evidence in which the forensic scientist had acted as the defence expert. For reasons of confidentiality it is not possible to state how many questionnaires were completed by each scientist.

Background to the case

The year in which the offences were committed is depicted by Table 4 below.

Table 4

Year	Number of Cases	
1987	4	(8.2%)
1988	13	(26.5%)
1989	10	(20.4%)
1990	10	(20.4%)
1991	7	(14.3%)
MISSING	5	(10.2%)
TOTAL	49	(100%)

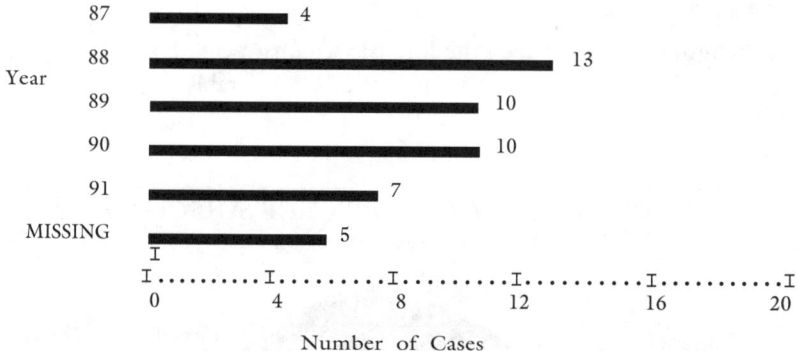

Number of Cases

The offence types covered by the cases dealt with in this questionnaire are depicted by the pie chart. Fig. 4.

In 38 cases (78 per cent) the defendant was charged with a sexual offence.

Figure 4.

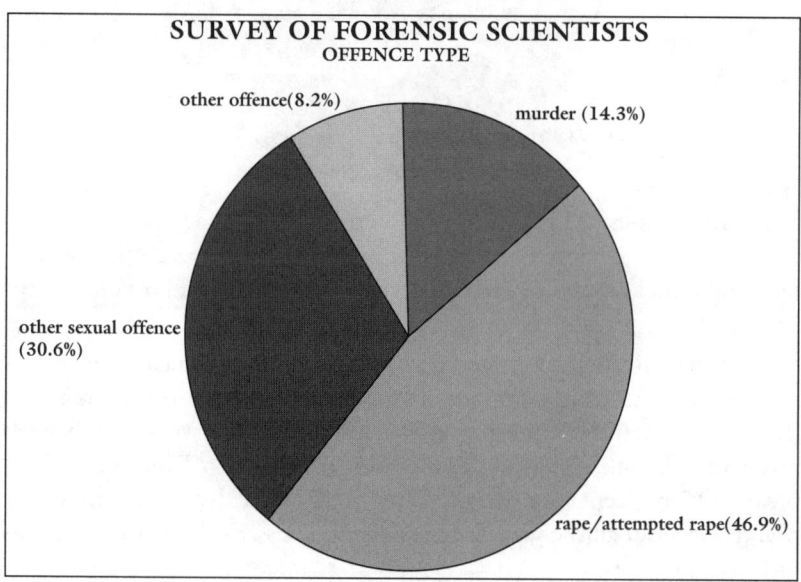

The forensic scientists were asked if they were aware of how the defence solicitor located them as an expert able to act in the case.

In 10 cases (20 per cent) the defence solicitor had been involved previously with the expert concerned.

In 6 cases (12 per cent) the defence solicitor located a suitable expert through another expert and in 4 cases the defence solicitor located the expert through information provided by the defence counsel or another solicitor. In 29 cases the experts were not aware of the means by which they were located.

Time available to Defence Expert

The length of time between the forensic scientist being contacted by the defence solicitor and the court date varied considerably. This is depicted by the pie chart Fig. 5.

Figure 5.

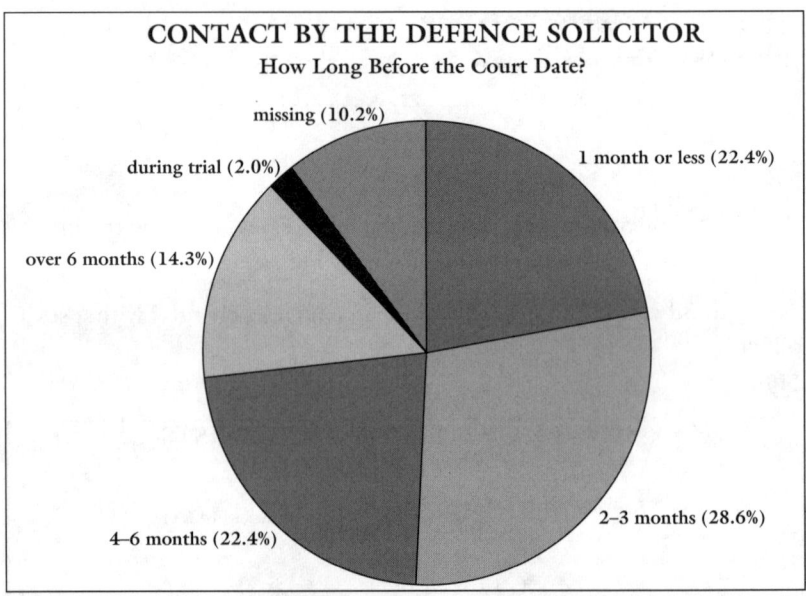

CONTACT BY THE DEFENCE SOLICITOR
How Long Before the Court Date?

missing (10.2%)

during trial (2.0%)

1 month or less (22.4%)

over 6 months (14.3%)

2–3 months (28.6%)

4–6 months (22.4%)

In 11 cases (22 per cent) the forensic scientist had 1 month or less pre-trial to carry out his work, and in 1 other case the expert was contacted during the trial. In 18 cases (37 per cent) the scientists had over 4 months to prepare.

It has already been stated that DNA profiling can take many weeks to carry out and so it is important to consider whether the scientist felt that they had sufficient time pre-trial to carry out all the work that they felt was necessary. The results of this question are shown in Table 5 below. In 6 cases (12 per cent) the defence expert had not had sufficient time pre-trial to carry out all the work he considered necessary. All 6 had 3 months or less to prepare.

Table 5:Were you notified in sufficient time to carry out all the procedures that you consider necessary?

Response	Number of responses
Yes	42 (85.7%)
No	6 (12.2%)
MISSING	1 (2.0%)
TOTAL	49 (100.0%)

In 7 cases (14 per cent) the forensic scientists stated that there had been time pre-trial to carry out the work but that it had been very rushed. In 3 of the 7 cases there was less than 1 month for the analysis.

Even where the defence only require evaluation of the prosecutions results or where this is the only available option this may take a considerable time. Facts to be taken into account here are the existing workload of the defence expert and that, for the independant evaluation, the defence expert will usually arrange to visit the prosecution laboratory at a mutually convenient time.

In view of these factors it is perhaps surprising that in 7 cases the defence expert was notified one month or less pre-trial and still managed to carry out the necessary work.

Legal Aid

In 47 cases (96 per cent) the work was funded by Legal Aid. In 10 cases (20 per cent) the defence expert considered that work was delayed whilst approval for Legal Aid was sought. See Table 6 below.

Table 6: Was the work delayed whilst approval for the Legal Aid was sought?

Response	Number of responses
Yes	10 (20.4%)
No	36 (73.5%)
MISSING	3 (6.1%)
TOTAL	49 (100%)

Table 7 below is a cross-tabulation of whether the defence expert was notified in sufficient time pre-trial and whether the expert felt that work was delayed whilst approval for Legal Aid was sought.

Table 7

	Was the work delayed whilst approval for Legal Aid was sought?		TOTAL
	YES	NO	
Were you notified in sufficient time to carry out all the procedures that you consider necessary?			
YES	9	31	40
NO	1	4	5
TOTAL	10	35	45

Surprisingly in the majority of cases where the expert felt the work had been delayed whilst the approval of Legal Aid was sought, they stated that there was sufficient time pre-trial for the work to be carried out. In only 1 of the 10 cases where there was a delay due to Legal Aid was there insufficient time to carry out the necessary work. There were 4 missing values to this table.

Work carried out by the Defence Expert

The defence expert was asked to indicate the type of work that he carried out on behalf of the defence. The respondents were free to reply to more than one category.

In 3 cases the defence expert repeated the analysis carried out by the prosecution.

In 5 cases the defence expert carried out laboratory investigations which differed from those carried out by the prosecution.

In 43 cases the defence expert carried out an independent evaluation of the results obtained, the laboratory records and the conclusion drawn by the prosecution. In 39 of these 43 cases this was the only method by which the defence expert evaluated the DNA evidence in the case.

Conclusions drawn by the Defence Expert

In 25 cases (51 per cent) the defence experts conclusions supported the prosecutions evidence in all respects.

In 21 cases (43 per cent) the defence experts conclusions differed from the prosecutions conclusions. In 2 of these cases the expert stated that the deviations were not significant. There were 3 missing values to this question.

The defence experts were invited to comment on why their conclusions deviated from those drawn by the prosecution experts. In 5 cases (24 per cent), where there was a difference of opinion, the defence expert disagreed with the band matching between the scene of crime sample and the defendants sample. In 10 cases (48 per cent), where there was a difference in opinion, the defence expert disagreed with the statistical calculations and/or database used by the prosecution expert. In 4 cases (19 per cent) the defence expert expressed concern over both the band matching and statistical inferences drawn by the prosecution.

Assistance provided by the Defence Expert

i. In 38 cases (78 per cent) the defence expert felt he provided information which assisted the defence at the trial e.g. in their cross-examination of the expert witness for the prosecution.

ii. In 10 cases (20 per cent) the defence expert attended court and advised the defence but was not called to give evidence.

iii. In 4 cases the defence expert attended court and gave evidence for the defence.

In 1 case another expert gave evidence in court for the defence.

In 7 cases the defence expert felt he had assisted in the manner described in both i. and ii. above. There were 4 missing values in this question.

Additional Comments by Defence Expert

In 2 cases the defence expert specifically commented that his report had supported the prosecutions conclusions and may have encouraged the defendant to plead guilty.

In 1 case the defence expert stated that the defendant had refused a police request for a blood sample for DNA profiling. However the forensic science laboratory agreed to a request by the police to carry out DNA profiling on a blood sample that had been provided by the defendant in connection with a previous offence.

CONCLUSIONS

DNA profiling is a complex scientific technique which has been used in forensic work since 1987 and was introduced into the courts comparatively early on in it's development. English law has no specific requirements regarding the admissibility of scientific evidence other than the fact that such evidence should be relevant and must not infringe any exclusionary rule. The scientific evidence is put before the court and it is for the jury, assisted by expert evidence, to assess the appropriate weight to give to the evidence. It is reasonable to argue that such an approach places a burden on the defence to disprove the evidence. Where such an approach is taken the Courts should ensure that the defence are able to obtain an independent assessment of the evidence and where appropriate challenge that evidence in court. The problems faced by defence lawyers will be of a similar nature irrespective of the particular forensic technique used. However each technique is different and the precise nature and extent of the problems may vary considerably.

This technique has been applied to forensic work since 1987 and the data available suggests that its use has increased rapidly over the last few years, and it is likely that it will continue to do so. There are a number of distinct methods of carrying out DNA profiling and the technique used in forensic casework has changed since its introduction in 1987.

DNA profiling is of particular value in certain offences against the person, both fatal and non-fatal, i.e. murder and sexual offences, particularly rape. It is in relation to these offences that a biological sample suitable for DNA profiling is likely to be left at the scene of the crime by the defendant, e.g. semen in an offence of rape. However another factor of influence in the type and number of cases where DNA profiling has been used is that it is an expensive and time consuming technique. The recent introduction by the Forensic Science Service of a form of DNA profiling[33] which is highly sensitive and comparatively quick to carry out, producing results in hours rather that days or weeks may mean greater use of this technique in any offence where an appropriate biological sample is recovered from the scene of the crime. In view of these factors, it is highly likely that the use of DNA profiling as a forensic technique is going to expand considerably in the near future. If this is allowed to happen then the courts must ensure adequate protection for the defendant. Many of the cases where DNA profiling has been used over the last 5 years have involved offences of murder or rape where the defendant may face a sentence of life imprisonment. In such circumstances it is vital that the courts should be aware of the problems faced by the defence in obtaining independent assessment of the evidence. These problems have been highlighted by the present research.

This research study aimed to determine some of the problems which face defence solicitors who wish to challenge DNA evidence. A number of interesting points have been raised by this study but it should be remembered that the research is based on a relatively small sample of completed questionnaires. Hence the research should be viewed essentially as an exploratory study, based on the perceptions of a small number of defence solicitors and experts.

In this study the first problem faced by the defence concerned disclosure of the existence of DNA evidence.

In the majority of cases the defence were notified by the prosecution, either CPS or police, of the existence of DNA profiling evidence. This result is not surprising in that it is the responsibility of the prosecution to inform the defence, as soon as is practicable after committal if they have not already done so, of the existence of expert evidence that they intend to adduce.[34]

The length of pre-trial notice that the defence had of the existence of DNA evidence varied enormously from 1 years notice to notification at the actual trial. In 10 cases (22 per cent) the defence lawyer felt that there was insufficient time pre-trial to obtain an independent assessment of the DNA evidence and in a further 4 cases the work was carried out but there was a considerable rush.

The opinion of the defence lawyers is supported by that of the forensic experts who acted for the defence. In 6 cases (12 per cent) the defence expert did not consider that he had sufficient time pre-trial to carry out the work he considered necessary and in 7 cases (14 per cent) the work was very rushed.

This situation is clearly unsatisfactory from the point of view of the defence. It may take a considerable period of time for the defence to locate a suitable expert and for the expert to carry out the required work. One reason for the late notification of the existence of DNA profiling evidence is that the prosecution may carry out DNA profiling work after the defendant has been charged and in some cases right up to the trial date. This occurred recently in the case of R. v Hassett (1992).[35] In one case in the study the prosecution applied for the trial to be adjourned in order that further DNA work be carried out. This application was refused.

It is apparent that disclosure of the evidence, to the defence, at the earliest opportunity is particularly important in relation to DNA evidence. It must be considered that the Crown Court Rules on disclosure of expert evidence are not specific enough and are inadequate in relation to DNA profiling evidence. The term "...as soon as practicable..."[36] allows the prosecution to carry out additional forensic work up to the trial date. It is suggested that where forensic work has been carried out after the committal date such evidence should not be admissible unless the defence have had adequate time to obtain an independent assessment of the DNA evidence. The judge may use his discretion under S78 of the Police and Criminal Evidence Act 1984 to exclude such evidence. A trial could be adjourned to allow for assessment of the DNA evidence by the defence, as occurred in 5 cases in this study. However it is reasonable to argue that the defendant who has been remanded in custody should not be prejudiced in this way. It is equally important to note that DNA work carried out after the committal date, such as the use of a different DNA probe, may result in the defendant being exculpated, so it is not suggested that such work by the prosecution should be prohibited.

Once the defence lawyer has learnt of the existence of DNA evidence in the case, he must then decide whether or not he wishes to consult an expert.

12 of the sample of 46 lawyers who responded to the questionnaire had not obtained independent assessment of the DNA evidence. There is no evidence from the results that any of the respondents to this category felt that the DNA evidence was overwhelming and that an independent assessment of such evidence was not worthwhile. That such a view might be held was suggested in an article on DNA profiling.[37] Moreover, defence lawyers who felt that the DNA evidence was otherwhelming may have been less likely to reply to the questionnaire.

All cases included in the sample were criminal cases and it was, therefore, not surprising that in the majority of cases the defence wished to apply for Legal Aid to cover the cost of their expert.

In 1 case in the study a defence lawyer indicated that an application for Legal Aid to cover the cost of DNA profiling work was refused. As stated earlier, DNA profiling has to date usually been carried out in relation to serious offences with the possibility of life imprisonment for the defendant who is found guilty. It is important in such a situation that the defence are given every opportunity to evaluate independently the prosecutions evidence. In these circumstances there should be no grounds for the refusal of Legal Aid. That such cases can arise has been indicated in the literature.[38]

The study of defence experts indicated that 47 of the 49 cases (96 per cent) were funded by Legal Aid. In 10 cases (20 per cent) the defence expert felt that the work was delayed whilst approval for Legal Aid was sought, although only one of the 10 said there was insufficient time to carry out the work. This highlights the problems for the defence in relation to the time available prior to the trial date. It is essential for the defence that a decision on the availability of Legal Aid for DNA work is made rapidly by the Legal Aid Board.

The defence may instruct a defence expert to begin work without prior authority from the Legal Aid Board in the hope that Legal Aid will be granted or that expenses may be recovered from Central Funds after the trial. However this is a risk to the defence solicitor who may incur considerable costs.

It is therefore proposed that where there is DNA evidence which implicates the defendant in a criminal offence, Legal Aid should be granted for an independent assessment of the DNA evidence, unless there are specific reasons for not doing so. The fact that the Legal Aid Board feel the DNA evidence against the defendant is overwhelming should not be good reason.

The granting of Legal Aid is particularly important in relation to a forensic technique which is as potentially discriminating as DNA profiling and which is still the subject of scientific dispute in the literature.[39]

The defence lawyer who wishes to obtain an evaluation of the DNA evidence must be able to locate a suitable expert who is able to carry out the work.

The defence lawyers who participated in the project located their expert in a variety of ways, many by word of mouth. This finding is supported by the responses of the defence experts themselves. All defence lawyers who took part in the study and who wished to locate an expert were able to do so, however in 3 cases the respondents experienced considerable difficulties in finding such an expert.

DNA profiling is a highly specialized technique and there are relatively few experts outside the FSS who are experienced in this field. Assessing forensic DNA profiling results is considerably different to assessing other forms of DNA profiling eg. paternity work, and it is therefore not surprising that there are a limited number of experts available to the defence.

Since the FSS adopted agency status in April 1991 it has carried out DNA work for the defence in a number of cases. That the experience of the FSS is recognized as available to the defence is very important. However, where the FSS have carried out the DNA work for the prosecution, the defence may be reluctant to employ them. It is therefore important to ensure access of the defence to other appropriately qualified experts.

The Law Society currently has a register of expert witnesses which includes a number of experts who are able to work in the DNA field. This register has been in existence for many years but has not been widely publicised. It is clear that a number of participants in the study were not aware of the existence of this register. The Law Society register has recently been updated and computerised and there are plans to publicise its existence later in the year.

In the study 9 of the 34 respondents who obtained an independent assessment consulted more than one expert. They may wish to do this for a variety of reasons but where the defence employ a number of experts purely in an attempt to find an expert who will disagree with the prosecutions conclusion the Legal Aid available should be limited. Legal Aid should be provided in all cases to employ the first expert but should only be provided for further experts where there is good reason. This would balance the need to be fair to the defendant with care in the way that Legal Aid is used.

The results suggest that in the majority of cases where an expert is employed by the defence, the work carried out is an evaluation of the prosecutions results. This will usually involve the defence expert visiting the laboratory where the forensic work was carried out in order to assess the conclusions drawn by the prosecution. The study of the defence experts suggests that in 80 per cent of cases this was the type of work that they carried out.

As discussed earlier, it is reasonable to argue that an evaluation of the prosecution's work with no laboratory analysis by the defence cannot amount to a fully independent assessment of the evidence. However this is often the only approach available to the defence expert due to insufficient crime stain remaining and/or insufficient time available pre-trial.

All the defence experts consulted during the course of this project stated that, where appropriate, they would prefer to carry out their own scientific analysis rather than simply evaluate the work carried out by the prosecution. However, where there is no scene of crime sample remaining, it is difficult to see how this problem can be overcome.

Due to the considerable time involved in carrying out DNA profiling it is impracticable to suggest that the prosecution's work be witnessed by a defence expert. This problem may to some extent resolve itself in the future as quicker and more sensitive techniques of DNA profiling are developed. For the present this is a serious problem faced by defence lawyers who wish to obtain a fully independent assessment of the evidence against their client.

In contrast to the view of defence lawyers and defence experts on this point, the MPFSL stated that they have never been asked by the defence to supply samples for independent analysis. It may be that there is in effect a lack of communication between the scientist for the prosecution and the defence. It may assist the defence if, as a matter of practice, the prosecution expert in his report included details of any material remaining after he has completed his work.

The amount of time available to the defence is again of importance here. Even where the defence wish solely to evaluate the results that the prosecution have obtained this may take several weeks. The defence must have sufficient time to locate a suitable expert. The number of experts is small and their workload high, it may be several weeks before the defence expert is able to arrange to visit the prosecution laboratory to examine the results. The defence experts consulted during the course of this study considered this to be a real problem and felt that they had turned down cases due to the lack of time available to carry out the work. This view was supported by a number of defence lawyers who had been forced to approach more that one expert as the first expert was too busy to carry out the work. This problem is exacerbated in cases where the defence wish to carry out laboratory work which may, with the current method of DNA profiling, take many weeks.

This study indicates that the defence lawyer who has been able to obtain an independent evaluation of the DNA evidence may find that the conclusions drawn by the expert differ from those drawn by the prosecution experts.

In the study 38 per cent of defence lawyers who had obtained an independent analysis of the evidence stated that their conclusions differed from those of the prosecutions' expert. The response from the defence experts themselves indicated that this occurred in 43 per cent of the cases that they had dealt with. In 3 cases these differences were stated to be insignificant, but this still indicates that in a proportion of cases the prosecution and defence experts disagree on the conclusions to be drawn from the evidence. This highlights the importance of the defence being in a position to obtain an independent assessment of the evidence. As stated in the results, this difference of opinion between prosecution and defence is reflected by the out-of-court arguments in the literature over the interpretation of the results of DNA profiling. It is noteworthy that the FSS have published a number of papers[40,41] concerning the method by which they interpret the results of DNA profiling evidence. It is important that such material is available to the defence experts.

The value to the defence of consulting an expert is clear. 94 per cent of defence lawyers who consulted an expert felt that they had been assisted by that expert, either in their evaluation of the case and the advice they gave to their client or in presenting their case in court.

22 of the 34 respondents who obtained an independent assessment wished to contest the prosecution's DNA evidence in Court. The defence lawyers required the assistance of their expert primarily to provide information in order to enable them competently to conduct this cross-examination or to attend court in order to advise on the cross-examination. In only 3 cases the defence called their expert to testify on their behalf. These results were reflected by the responses of the defence experts.

Where such circumstances exist, this cannot be a waste of Legal Aid money and it is reasonable to argue that the defence will be disadvantaged if they are unable to employ their own expert.

It is also significant that the primary role of the defence expert appears to be that of providing information to enable the defence to conduct their cross-examination of the prosecution expert. In only a small number of cases was the expert called to give evidence. It is postulated that the reason behind this is that although there is dispute over the interpretation of DNA results and in particular the calculation of the statistics, the basic technique is not disputed.

Of the defence lawyers who obtained an independent analysis of the DNA evidence 47 per cent reported having experienced specific problems concerning the DNA evidence. If DNA profiling continues to be used as a forensic technique it is essential that these problems be addressed, whatever system, adversarial or inquisitorial, operates.

The comments made by the individual defence lawyers indicated once again the problems in relation to disclosure, Legal Aid and locating a suitable expert. It is reasonable to expect that such problems occur in relation to many forensic techniques but, as has been discussed above, such problems are highlighted by an expensive, time consuming and potentially highly discriminatory technique such as DNA profiling.

In English Criminal Law the approach to scientific evidence is based on relevance. There is no requirement for the scientific evidence to have been accepted as reliable by the relevant scientific community or for there to have been papers published in relevant journals on the technique. Some states in America adopt such an approach termed the Frye Standard[42] which has been criticized as unduly restrictive and expensive with long pre-trial hearings on the validity of a technique. If English Law wishes to avoid such an approach, awareness of the problems faced by the defence is of prime importance in considering whether or not evidence resulting from a particular scientific technique should be admissible. It is proposed that scientific evidence should not be admissible against a defendant unless there is within the scientific community an independent expert who is knowledgable of that technique and is able to evaluate the conclusions drawn by the prosecution. This will infer a knowledge of a particular technique within the relevant scientific community. This may raise specific problems in relation to DNA profiling, which is perhaps unique in the speed at which the technique has developed since its introduction into forensic case work. Nevertheless evidence produced by the prosecution which involves the use of a new DNA probe in the traditional methods of DNA analysis or the use of a new method of DNA profiling should be approached with extreme caution. The courts should ensure that the defence had an opportunity to consult an expert who has knowledge of that probe or technique and has been able to assess the evidence for the defence. Comments received in the study indicate that situations have arisen where the defence have not had access to the exact materials used by the prosecution, eg. DNA probes, for commercial/ copyright reasons. It is unacceptable that the defence be prejudiced in this way. Where such material is withheld from the defence the prosecution's analysis should be inadmissible.

The research confirms that there is scope for dispute over the interpretation of the results of DNA profiling[43] albeit that the basic technique is accepted worldwide[44] and seen as being of considerable value when identification is in issue. It is therefore important, if DNA profiling continues to be used in forensic casework and is allowed to develop further, that the problems which face defence lawyers when dealing with this technique are seen to be addressed. Of crucial importance is the ability of the defence to obtain an independent evaluation of the prosecutions evidence.

The courts must ensure that the defence are notified of the existence of DNA evidence in sufficient time pre-trial to enable them to locate a suitable expert and for that expert to carry out all the necessary work. Legal Aid should be granted automatically for one expert assessment of the prosecution work. DNA evidence should only be admissible where an appropriate expert is available to the defence.

If the defence are able to show on the balance of probabilities, that the above criteria have not been met the prosecution evidence will be unduly prejudicial and should be excluded.

REFERENCES

1. DNA – Deoxyribonucleic Acid – the genetic material for higher organisms.
2. **Jeffreys A.J. Wilson V., and Thein S.L.** *Nature* (1985) v 314 pp. 67–73.
3. Multilocus Probing (MLP) – the use of a genetic probe which identifies a number of sites of DNA variability simultaneously so producing a pattern of many bands. This technique has been termed DNA or genetic fingerprinting.
4. **Werrett D., and Lygo J.** *The Law Society's Gazette* (1987) pp. 3637–3639.
5. **Sensabaugh G.F.** *Journal of Forensic Science* (1986) pp. 393–396.
6. **Gill P., Jeffreys A.J. and Werrett D.J.** *Nature* (1985) v 318 pp. 577–579.
7. **Gill. P., Lygo J.E., Fowler S.J., and Werrett D.J.,** *Electrophoresis* (1987) v 8 pp. 38–44
8. Single Locus Probing (SLP) the use of a genetic probe which identifies one specific area of DNA variability are typically produces a pattern of two bands one inherited from each parent.
9. **Lygo J.** *New Law Journal* (1991) v 141 No. 6498 pp. 448–452.
10. **Jeffreys A.J., Macleod A., Tamaki K., Neil D.L., and Monckton D.G.** *Nature* (1991) v 354 pp. 204–209.
11. *Police Review* August 16th 1991 p. 1660.
12. R v Melias (1987) *Times* 14th November 1987.
13. *Times* 14th November 1987.
14. People v Joseph Castro, Bronx County, New York, Criminal Term Part 28 1989.
15. R v Tran (1991) *Criminal Law Review* 1991 pp. 583–590.
16. People v Randolph Jakobetz (1992) United States Court of Appeals for the second circuit.
17. News and Comment Science 1991 Vol. 254 pp.1721–1723.
18. **Chakraborty R., and Kidd K.** *Science* 1991 Vol 254 pp.1735–1739.
19. **Lewontin R.C., and Hartl D.L.** *Science* 1991 Vol 254 pp.1745–1750.
20. *The Sunday Times* 29th December 1991 p.4.
21. *The Guardian* Wednesday 8th January 1992.
22. Data provided by Dr D.J. Werrett, Forensic Science Service.

23. Information provided by Dr Peter Martin, Metropolitan Police Forensic Science Laboratory.

24. *Law Society's Guardian Gazette* Post Box March 4th 1992.

25. op. cit. n.11.

26. **McLeod N.,** *Criminal Law Review* (1991) pp. 583–590.

27. See page 5.

28. Electrophoretic gels – the gels which are used to sort and retain the DNA sections according to size; thereby retaining part of the sample.

29. Autoradiograph – developed X-ray film showing distances migrated by the various fragments of DNA. Usually termed the DNA profile.

30. op. cit. n.9.

31. **I.W. Evett and R. Rinchin.** *Int. J. Leg. Med* (1991) 104 pp. 267–272.

32. op. cit. n.17–19.

33. op. cit. n.11.

34. Crown Court (Advance Notice of Expert Evidence) Rules S1 1987 No. 716.

35. R. v Hassett (1992) Birmingham Crown Court. March 1992.

36. Crown Court (Advance Notice of Expert Evidence) Rules S1 1987 No. 716.

37. **Andrew Hall** *New Law Journal* (1990) v 140 pp. 203–205.

38. **Russell Stockdale** *New Law Journal* (1991) v 141 p 772.

39. op. cit. n.17–21.

40. op. cit. n.31.

41. **I.W. Evett and P. Gill** *Electrophoresis* 1991 12 pp. 226–230.

42. Frye v United States 293 F1012 (D.C. Cir 1923).

43. op. cit n.17–19.

44. op. cit. n.16.

APPENDIX 1

QUESTIONNAIRE

This questionnaire is aimed at evaluating the ability of defence lawyers to challenge DNA profiling evidence, particularly in relation to economic constraints and the avail-ability of relevant expert witnesses.

Please complete a questionnaire for each case you have dealt with involving DNA profiling evidence, whether or not you sought expert assessment of the evidence and whether or not the DNA evidence was significant in the case. Please add additional comments to the sheet where you feel this would help to explain the circumstances.

PLEASE LEAVE BLANK — FOR OFFICE USE ONLY

A   Background the case

1. In what capacity were you involved in this case:

    _____

    _____

    _____

2. Was this the first case that you had dealt with involving DNA profiling evidence?

    NO   _____

    YES   _____

3. Offence type: _____

4. Year the alleged offence was committed   _____

5. Did defendant plead

    GUILTY   _____

    NOT GUILTY   _____

6. Outcome of the case –

    _____

    _____

    _____

B Disclosure of DNA profiling evidence

1. Who notified you of the existence of DNA profiling evidence case?

   _____

2. When were you notified, in relation to the trial date, of the existence of DNA evidence?

   _____

3. Was there sufficient time pre-trial to obtain your own expert assessment of DNA evidence had you so wished?

   NO   _____

   YES   _____

Additional comments-:

   _____
   _____
   _____

C Expert assessment of the DNA evidence for the Defence

1. Did you wish to obtain an expert assessment of the DNA evidence on behalf of your client?

   NO   _____

   YES   _____

   Please clarify your answer by replying to question 2 or 3, whichever is appropriate.

2. You did *not* wish to obtain an expert assessment of the DNA evidence on behalf of your client.

   a) The DNA evidence did not implicate your client.   _____

   b) You did not feel it was necessary to obtain an expert assessment of the DNA evidence. Your client did not admit the offence but offered an alternative explanation e.g. admitting his presence at the scene but denying the offence, claiming consent to an allegation of rape.   _____

   c) The client decided to plead guilty prior to being aware of the DNA evidence against him.   _____

PLEASE LEAVE BLANK

d) The client decided to plead guilty on becoming
   aware of the DNA evidence against him. _____

e) The DNA evidence was felt to be overwhelming
   and an expert assessment of the evidence was not
   considered necessary. _____

f) Other reason – please specify

   _____

   _____

   Comments to clarify the situation if appropriate —

   _____

   _____

   _____

3. You *did* wish to obtain an expert assessment of the
   DNA evidence

   a) Did you apply for legal aid to cover the costs?

      YES   _____

      NO   _____

   b) Was Legal Aid granted?

      YES   _____

      NO   _____

   c) Did you approach an expert for an assessment of
      the DNA evidence:

      YES   _____

      NO   _____

   d) How did you locate a suitable expert?

      _____

      _____

      _____

   e) How many experts did you approach? _____

f) If you approached more than 1 expert what assistance did each expert provide?

_____

_____

g) Did you instruct an expert to carry out an assessment of the DNA evidence on your behalf?

      YES   _____

      NO   _____

h) If you are aware of the type of work carried out by the expert please tick one or more of the below as appropriate

   i)   a repetition of the investigations carried out by the prosecution   _____

   ii) laboratory investigations which differed from those carried out by the prosecution   _____

   iii) independent evaluation of the results obtained, the laboratory records, and the conclusions drawn by the prosecution   _____

D Results of the defence experts assessment

1. The experts analysis:–

supported the prosecution's conclusions in all respects   _____

deviated from the proseuction's conclusions   _____

Where the assessment deviated from prosecution's conclusions, if you are aware of details, i.e. in relation to scientific process, statistics, etc, please specify –

_____

_____

_____

50

E Challenging the DNA Evidence in court

1. Did you wish to challenge the DNA evidence in court?

        YES   _____

        NO   _____

2. Did you require the assistance of the expert witness

    a) to provide you with *information* in order that the defence could cross-examine the expert witness for the prosecution?   _____

    b) to *attend* court in order to *advise* the defence on cross examination of the expert witness for the prosecution?   _____

    c) to *attend* court to give *evidence for the* defence?   _____

3. Did you apply for Legal Aid to cover the costs of an expert attending court?

        YES   _____

        NO   _____

4. Was Legal Aid granted?

        YES   _____

        NO   _____

5. Was the expert witness called by you to give evidence?

        YES   _____

        NO   _____

6. Outcome of case:–

_____

_____

_____

7. Do you consider that consulting an expert witness assisted you in assessing the case and choosing what course to take or assisted you in court in the defence of your client?

YES _____

NO _____

Please give your reasons briefly –

_____

_____

_____

_____

8. Please specify any particular problems you experienced in relation to DNA evidence in this case.

_____

_____

_____

_____

APPENDIX 2

QUESTIONNAIRE

This questionnaire is aimed at collating information on criminal cases where DNA profiling evidence has been available. It is designed to be completed by forensic scientists who have acted in such cases, and it will complement a questionnaire sent to defence lawyers. Please complete a questionnaire for each case and feel free to add additional comments where you feel that this would clarify the circumstances.

1. Year the offence was committed   _____

2. Offence type _____

3. How did the defence solicitor locate you as an expert able to act in cases involving DNA evidence?

    _____

    _____

4. How long before the court date were you contacted by the defence solicitor?   _____

5. Were you notified in sufficient time to carry out all the procedures that you consider necessary?

             YES   _____

             NO   _____

    Additional comments –

    _____

    _____

    _____

    _____

6. Was the work funded by Legal Aid?

        YES   _____

        NO   _____

7. Was the work delayed whilst approval for the Legal Aid was sought?

        YES   _____

        NO   _____

8. Please tick one or more of the below to indicate the type of work carried out

    i)   a repetition of the investigations carried out by the prosecution   _____

    ii)   laboratory investigations which differed from those carried out by the prosecution   _____

    iii) independent evaluation of the results obtained, the laboratory records and the conclusions drawn by the prosecution   _____

9. Did your assessment of the evidence –

    i)   support the prosecution's evidence in all respect   _____

    ii)   deviate from the prosecution's conclusions   _____

Where the assessment deviated from the prosecution's, please clarify e.g. number of bands visible.

_____

_____

_____

_____

_____

10. In your capacity as an expert witness which of the
    following applied in this particular case —

    Please tick

    i) you were not able to be of assistance to
       the defence   _____

    ii) you provided *information* which assisted
        the defence at the trial, e.g. in their cross-
        examination of the expert witness for the
        prosecution   _____

    iii) you *attended* court in order to *advise* the
         defence but were not called to give
         evidence   _____

    iii) you *attended* court and gave *evidence* for
         the defence   _____

    Additional comments –

    _____

    _____

    _____

    _____

    _____

Printed in the United Kingdom for HMSO.
Dd.0297202, 1/93, C6, 3396/4, 5673, 221519.